Kleydiane Silva

Didactic games: a form of intervention in maths lessons

Kleydiane Silva

Didactic games: a form of intervention in maths lessons

Using play as a teaching strategy

ScienciaScripts

Imprint

Cover image: www.ingimage.com

This book is a translation from the original published under ISBN 978-613-9-66914-1.

Publisher:
Sciencia Scripts
is a trademark of
Dodo Books Indian Ocean Ltd. and OmniScriptum S.R.L publishing group

120 High Road, East Finchley, London, N2 9ED, United Kingdom
Str. Armeneasca 28/1, office 1, Chisinau MD-2012, Republic of Moldova, Europe
Printed at: see last page
ISBN: 978-620-8-11212-7

Dedication

To my God, the author and finisher of my faith, who made me overcome every obstacle by giving me the strength and wisdom to get this far.

To my parents, Luiz Tiago de Sousa Carvalho and Maria da Glória Silva da Costa de Sousa, for making my dreams their own.

To my husband Carlos Augusto da Silva for his support and encouragement.

.

Kleydiane Silva de Sousa

Summary

INTRODUCTION

This book is an exposition of studies and classroom experiences on mathematical games as a teaching-learning methodology in maths classes, which has been going on since our first contacts in the classroom, which began in our undergraduate internships.

When we started our undergraduate studies, we always wanted to graduate and be educators who would make a difference, unlike many of the maths teachers we had throughout our student life, because although we had a facility with numbers, we felt indignant about the way our classes were presented, always a sameness in which most students never understood the real reason for studying maths.

In recent years, more and more studies have been carried out to try to alleviate the huge deficit in maths teaching. These studies point to methodology as the root cause of this "deficit", as well as issues such as teacher training, inadequate textbooks, lack of resources, planning, programme content, among others.

The way students see maths will only change if we also change the way we see the teaching of maths, because as D'Ambrósio (1991, p.1) says "(...) there is something wrong with the maths we are teaching. The content we are trying to pass on through the school systems is outdated, uninteresting and useless." In a world where technological advances are attracting young people every day, boring lessons, where students don't see any association between the content taught and their daily lives, become less attractive, which contributes to this uselessness and lack of interest on the part of students in teaching maths.

Lessons that are always monotonous and expository, without detracting from expository lessons, because they are fundamental for the definitions of the theoretical framework, should not be the only methodology to be used in the classroom, especially in maths lessons, where students need to be motivated to better understand the essence.

This book was motivated by the search for a proposal that would collaborate, or why not say, help the maths teacher with a methodology that seeks to motivate students to take an interest and develop skills, leading them to get to know the world of maths that is found in their daily lives in a pleasurable way. We are not claiming that this proposal will solve all the problems that have led to deficiencies in maths teaching, nor are we looking for a formula to be followed so that students succeed in teaching maths. Our aim is to show that games are a pedagogical tool that can help mediate the maths teaching-learning process, seeking to attract students in a way that challenges them, gives them pleasure and instils an interest in the mother of all sciences.

Motivation is the shot in the arm that students need when it comes to teaching maths, because we know that there is a huge deficit when it comes to teaching this subject right from the early grades. Students feel demotivated because they often don't understand the application of this subject and its contents in their daily lives. This decontextualisation is often due to the lack of an appropriate or motivating methodology to arouse students' interest and integrate them into their studies.

Mathematical games are a very complete methodology to help in maths classes, as they integrate, motivate and develop skills and competences in students that allow them to develop their logical reasoning, not to mention that the student learns by playing and having fun, in other words, the student socialises and at the same time develops skills that are essential to the teaching-learning process in a pleasurable way.

The use of problem-solving games is a valuable tool to help maths teachers in their classes, so that they can interact and encourage students to develop the strategic skills and competences that are fundamental for intellectual development, which is essential for learning maths. Not to mention that games are always motivating, as they develop the competitive and strategic spirit of human beings, regardless of their age group.

With the aim of verifying whether the use of mathematical games contributes to improving student performance in upper primary and secondary school maths classes, we tried to carry out interventions in maths classes in the most diverse years of basic education, from 6th grade to 9th grade and from 1st grade to 4th grade, in the various types of basic education. We proposed the use of games alongside mathematical content, and these experiments took place in state and municipal schools in the city of Floriano-PI, between 2011 and 2014.

This book has been divided into chapters, which initially provide an overview of the theoretical framework and discussions on the difficulties encountered by teachers and students in the process of teaching and learning maths in the classroom, followed by some experiences with games in the classroom at various levels of basic education. We hope that this work will contribute to the enrichment and reflection of educators and teachers in basic education, which is extremely important for the good performance and training of citizens.

CHAPTER 1

EDUCATIONAL GAMES THROUGHOUT HISTORY

Since the dawn of humankind, there have been ancient manuscripts reporting the existence of games and some scholars have commented that along with this emergence came the need to play. There are cave paintings and archaeological remains dating back to BC that show that the Greeks and Romans played games. It is not known for sure which was the first game to be created, and there are disagreements between supporters of Darwinian theory and researchers of Christian doctrine.

Great names in history have stood out as great players, among them Joan of Arc, King Arthur, Theseus, Galileo Galilei, Julius Caesar, Napoleon Bonaparte and others. All these great names have contributed to the enrichment and creation of many games, some of which do not exist today and others which have been perfected.

Games have always been part of life in society and from the beginning they were used as leisure and as a tool for generating culture. Kishimoto (2001) mentions this importance when he says that it is essential for the development of social life, because when played in groups it develops interactivity and critical thinking, in short, human beings learn rules of behaviour and living through games, integrating themselves into different social classes.

Murcia (2005) describes the intimate bond between play and the human species. He says that play has always existed and that this activity is as old as humanity. He says that human beings have always played, sometimes more, sometimes less, and that it is through play that man has learnt to live and even dares to say that a people's identity is linked to play.

The vision of play as a teaching strategy, which is the aim of this work, emerged with Plato. He was one of the first scholars to have the vision of using play for educational purposes when he pronounced the importance of "learning through play". Aristotle was also one of the forerunners in the use of games to teach, and it was he who suggested games for the education of young children, with the aim of preparing them for life. But it wasn't until the middle of the 17th century, with the ideals of the humanists, that play began to be seen as an "educational resource", at first for reading and calculus, and then expanding to other areas of science.

In the mid-18th century, innovative games were created to teach science to royalty and the aristocracy. It was only after some time that the games became popularised, but with the aim of indoctrinating the population, as the games recounted historical events, criticisms and ideals that were established by the nobility. Thus, games that had previously been restricted to the education of the nobility were introduced to the population as an instrument for propagating ideals.

It wasn't until the 19th century, with the end of the French Revolution, that games began to be produced and seen as educational material. Although there has always been a link between play and learning. Rousseau (17121778), Pestalozzi (1746-1827) and Froebel (1782-1852) were big names who defended the importance of play for children's development, i.e. using it as an aid in the teaching-learning process.

The spread of play, as well as its influence on the teaching-learning process, only culminated in the middle of the 20th century through research and the formulation of theories aimed at elucidating this relationship between play, knowledge and development. Decroly (1871-1942), with the creation of materials aimed at the education of mentally handicapped children, and Maria Montessori (1870-1952), with her pedagogical ideals, were big names that gained acceptance in Brazil and around the world when referring to this theme. Bruner, Vygotsky and Piaget are other great 20th century researchers who have stood out by defending play as a pedagogical tool for children's development and learning.

But it wasn't until the 1980s that educational games gained momentum in Brazil. This was thanks to the emergence of toy libraries, the growth of congresses and the increase in scientific productions on this subject. Despite arriving with great force, this pedagogical tool is still little used by educators in the Brazilian public school system.

CHAPTER 2

THE GAME AS A DIDACTIC-PEDAGOGICAL TOOL IN THE CLASSROOM

As we reported in the previous chapter, the game is a cultural element that is closely linked to the human species. It is therefore a tool that has many attributes for the development of man, as a species that is constantly adapting to the environment, proving to be a powerful weapon to aid the teaching-learning process in the classroom.

The educational game has two completely antagonistic functions: the ludic function and the educational function. The ludic function refers to the pleasurable, spontaneous, fun and discovery part, while the educational function is related to the acquisition of knowledge and the complementing of knowledge. When these two functions are balanced, they fulfil the main objective of the educational game, but when this doesn't happen, the game is just another game, where there is no teaching or acquisition of knowledge. But how do we achieve this balance between these two functions? This task falls to the teacher, and we'll talk about this in the next chapter.

It is well known that Didactics is the art of teaching and that its field is very broad and complex, as it involves various subjects and elements of the educational process: teachers, students, content, methods, educational objectives, among other factors both internal and external to the teaching process. But we mustn't forget that its main objective is to look for ways to facilitate and collaborate with the teaching-learning process. The pedagogical tool fulfils this objective, as it is an instrument that contributes to facilitating the student's learning. In other words, a didactic-pedagogical tool is nothing more than an aid that helps both the teacher in the teaching process and the student in the learning process.

When the game is used in the classroom as intervention material, it assumes its pedagogical function, but this requires organisation, defining goals to be met and objectives to be achieved. If this is not well defined, the game will become just another activity to fill class time, adding no value to the student's life. In order for the game to fulfil its didactic-pedagogical function, it needs to be seen as an instrument for research and teaching by the educator, and learning and acquiring knowledge by the student.

The use of games in the classroom is defended by many scholars and researchers in the field of education due to the fact that the game is a very complete activity, as it encourages students to get involved in a pleasurable way, rescuing the investigative, creative and competitive spirit, releasing in students the desire to learn and know more about something new, taking them out of the routine classroom environment, with boring lessons and uninteresting exercises that make no sense to their lives, causing them to lose interest, curiosity and the pleasure of studying.

Learning through games provides access to interesting and fun learning. However, they should be adopted at random to remedy some of the shortcomings that occur in the classroom on a daily basis. There are three aspects that alone justify the inclusion of games in the classroom: the playful function, intellectual development and the formation of social relationships.

Jean Piaget (1896-1980) was a biologist and philosopher who dedicated himself to studying the process of developing an individual's intelligence. He was one of the great cognitivist theorists, where this current of learning argued that biological maturation, prior knowledge, language development, the process of social interaction and the discovery of affectivity are important factors for intellectual development and, consequently, learning. Piaget was a great advocate of the practice of games for the intellectual and social development of human beings.

Piagetian theory points to four factors responsible for a child's cognitive development:

- The biological factor: an individual can only learn a particular piece of knowledge if they are intellectually mature and thus prepared to receive it;
- The exercises and experiences acquired through the child's manipulation of the object;
- The social interactions that occur through language and education;
- The balancing factor of actions that encourages children to find answers to new problems.

In the process of development and learning, the child is seen as an active agent who, as they interact with the world around them, begins to modify their reality. Acting, in the Piagetian sense, is not just external and visible movements, but also internal, cognitive and affective activities. For example, in a game where the child compares, sorts, classifies, counts or makes mental deductions in order to solve it, they are mentally active.

According to Piaget's theories, the child's actions depend on their action schema. It is through this schema that they interpret and organise their action so that it can be put into practice, i.e. in different situations they make small modifications so that the new objective can be achieved. When faced with a new game, the child stimulates their existing action schemas and modifies them to suit the new conditions.

Jean Piaget's Psychogenetic Constructivism argues that the child's intellectual development is in a constant process of balancing with the environment. Thus, every time the child encounters a new situation, an imbalance is created. The child then looks for new schemes or ways to adapt to this new situation and return to a state of equilibrium. When the child's intellectual structure is not sufficiently developed to deal with this new situation, what Piaget calls major equilibration takes place. This process involves two mechanisms, assimilation and accommodation, which allow new schemes to be built and consequently intelligence to be developed. Assimilation is the addition of new knowledge to the intellectual structure that is not modified, while accommodation is the reorganisation of the mental structure so that this new knowledge, information or experience can be incorporated, adjusting to the demands of the environment. When accommodation occurs, the child returns to a higher state of equilibrium.

Lev Semyynovich Vygotsky (1896-1934), a researcher who was a contemporary of Piaget, developed the Sociointeractionist Theory in which he emphasised the contributions of culture, social interaction and language to the process of development and the subject's historical social learning. This theory emphasises the factors present in play as being essential for human development and learning.

Jerome Bruner (1915-2000) was a great cognitivist psychologist who developed the Theory of Instruction in the 1980s in the United States. This theory values social interaction and dialogue in the learning process. It is based on fundamental principles that are present in the art of play, namely: motivation, structure, sequence and reinforcement. Motivation is linked to the fact that every child has the desire to learn, i.e. motivation is an intrinsic factor that predisposes learning. Structure refers to the way in which the content to be transmitted is organised, so any content can be transmitted and understood, it all depends on how it is organised (or structured) and adapted to the recipient's intellectual conditions. The sequence used to present the content will determine how difficult or easy it is for the student to understand what is being presented. And last but not least is reinforcement, which is the feedback that lets the student know how they are doing.

A great psychologist and pedagogue, David Ausubel was a great advocate of cognitivist theories and in 1985 created the Theory of Meaningful Learning. This theory states that we relate new content, ideas or information to existing concepts in our cognitive structure, which we call anchor points for learning. Thus, for learning to be meaningful, the material to be assimilated by the student must be significant enough for them to be able to establish anchor points.

From what has been said about cognitivist theories, we can see that games are a didactic-pedagogical tool with qualities that are fundamental for meaningful learning, provided they are planned and adapted according to the intellectual needs of each age group of students. In this way, lessons can become more attractive and productive, as long as there is an organised sequence of content and activities to be carried out.

In short, there are many benefits that games bring to human beings, as they develop the physical, psychological, affective, motor, moral and emotional aspects. It's a very complete activity that, when used for educational purposes, develops competences and skills that facilitate the teaching-learning process in the classroom, especially with regard to mathematical content.

CHAPTER 3

THE USE OF GAMES IN MATHS TEACHING

It's not that long ago that Brazilian schools, with a few exceptions, adopted the concept that in order to be considered intelligent, a person had to have linguistic and mathematical skills, and the latter stood out even more. The other skills were left aside or disintegrated from these two skills, which are undoubtedly the basis for all the others.

The way maths is taught in state schools here in Brazil has not helped to show students the importance of this subject in their social, cultural and political lives. Tedious and monotonous lessons, repetitive and even exhausting exercises, a veritable exhibition of ready-made knowledge, thus creating an unsuitable environment for the teaching-learning process of this subject. Lara (2003) mentions the crisis in maths teaching and blames the problems on methodology, teacher training, inadequate textbooks, lack of resources and programme content. The author believes that it is up to teachers to rescue the student's desire to learn maths.

> "If we consider that teaching maths means developing logical reasoning, stimulating independent thinking, developing creativity, developing the ability to handle real situations and solve different types of problems, we will certainly have to start looking for alternative strategies." (LARA, 2003, p. 21)

Students need to see maths as something that is part of their reality and that it is possible to learn in a joyful and dynamic way. We admit that maths teaching needs to be changed, making it useful in the student's life and introducing them to the labour market. This change will only be possible with the use of different methodologies that motivate students to want to take part in the universe that is maths. According to Lara (2003),

> "This bogeyman or terror of our students will only lose its aura of the 'big bad wolf' when we, as educators, focus all our efforts on ensuring that teaching maths is about: developing logical reasoning and not just copying or exhaustively repeating standard exercises; stimulating independent thinking and not just mnemonic ability; developing

> creativity and not just transmitting ready-made knowledge; developing the ability to handle real situations and solve different types of problems and not continuing in that 'sameness' that we experienced when we were pupils." (LARA, 2003, p.18-19)

The high failure rate in maths is something that worries teachers of this subject. It is notorious that even students who pass this subject often do not know where and how to use the information they have learnt in this subject. Fiorentini and Miorin (1990) report on this:

> "The difficulties encountered by students and teachers in the maths teaching-learning process are many and well-known. On the one hand, students are unable to understand the maths taught to them at school, they often fail this subject, or even if they pass, they find it difficult to use the knowledge they have 'acquired'; in short, they are unable to effectively access this fundamentally important knowledge." (p. 5).

The arduous task lies with the teacher, whose main objective is to provide the student with a favourable environment for their development and learning. Alves (2009) discusses the difficult task of teaching maths and criticises the type of methodology used in most schools today, especially in maths classes:

> "It was because I believed that educators should be able to create an atmosphere of interest and motivation in the classroom, allowing students to participate fully and autonomously in the teaching-learning-evaluation process, that I didn't want to be a mere repeater of content, but to maintain a dynamic approach to its application." (p. 11)

An abrupt change in the methodology used in maths lessons is essential. The practice of traditional pedagogy, where the teacher is the central figure and the student is just a passive subject, blocks the student's interest, making it impossible for them to enjoy classes, attend them and even study this beautiful and challenging subject. It is a challenge for the educator to propose innovative methodologies that will rescue the student's desire to learn maths and reduce their traumas and fears in relation to this subject.

There are currently a number of innovative and technological methodologies that can be used in the classroom to help the teacher offer a more interesting and exciting lesson, arousing the student's curiosity to learn. These include: manipulative materials, computers, educational games, interactive whiteboards and others. We chose to work with games because we believe that they are one of the most complete methodological tools, based on studies in educational psychology, and because we believe that there are more possibilities for attracting the student's interest since they have playfulness in their favour.

Murcia (2005) defends play as a healthy and enjoyable educational activity that contributes to the student's development in the educational process:

> "The characteristics of play make it an ideal learning and communication vehicle for the development of the child's personality and emotional intelligence. Having fun while learning and getting involved with learning make the child grow, change and actively participate in the educational process." (p. 10).

There are numerous reasons for using games as a teaching tool to help teachers in maths lessons:

> "Another reason for introducing games into lessons is the possibility of reducing the blockages presented by many of our students, who fear maths and feel incapable of learning it. Within the game situation, a passive attitude is impossible. We've noticed that while these students play, they show better performance and positive attitudes towards their learning processes." (BORIN, 2002, p.9)

The National Curriculum Parameters (PCN's) defend the use of games in maths lessons, dismissing important characteristics for student learning through this resource:

> "Games are an interesting way of posing problems, as they allow them to be presented in an attractive way and favour creativity in the development of resolution strategies and the search for solutions. They allow for the simulation of problem situations that require lively and immediate solutions, which stimulates the planning of actions; they make it possible to create a positive attitude towards mistakes, since the situations happen quickly and can be corrected naturally, during the course of the action, without leaving negative marks.""(MEC, 1998, p.47)

The PCN's harshly criticise maths teachers, blaming them for the failure of their students. Mechanised lessons, where the content is merely a reproduction or accumulation of information that has no relation to the students' daily lives, are not at all attractive for students to awaken their investigative spirit and for teachers to fulfil their role as mediators of knowledge. It is necessary to use more attractive methodologies or teaching resources in order to reduce the distance between students and maths. We mustn't forget the importance that the PCN's have nationally in the political pedagogical project and in the activities that should be carried out in the classroom by maths teachers.

As we mentioned in the previous chapter, Piaget was a great advocate of using games for children's intellectual development, because even mistakes can lead to intellectual autonomy. He criticises the teaching of traditional mathematics, because according to him the relationship is, "one that links a teacher, a kind of absolute sovereign, holder of intellectual and moral truth, to each pupil considered individually." (Piaget, 1975), and although the content taught is modern, the way of teaching remains archaic. He argues that every student is capable of learning,

> "Every normal pupil is capable of good mathematical reasoning as long as his activity is appealed to and it is thus possible to remove the affective inhibitions which quite often give him a feeling of inferiority in lessons dealing with this subject." (PIAGET, 1975, p.65).

In this book, we are not saying that exercises and lectures are unnecessary, and that only games are the methodology that should be adopted in maths or any other subject, but that the application of games in the classroom, and in particular in maths lessons, is a motivating methodology that develops skills and competences in students and instils in them a desire to see this subject in a more interesting and attractive way. We support the idea of Kammi and Declark (1992), who state that "children are more active when they play what they have chosen and what interests them than when they fill in exercise sheets." (p.172).

The use of games as a teaching-learning strategy in maths classes has been debated and cited in many maths teaching studies. This tool provides a more interesting and motivating environment for the teaching-learning process, making lessons more dynamic and less traumatic. This playful activity makes the classroom environment more productive and enjoyable for both the teacher and the student. And we continue to emphasise that we are not trying to replace activities, but to propose a new one to help in the educational process of developing the student. As Petty (1995) states:

> "Play is one of the activities in which children can act and produce their own knowledge. However, our proposal is not to replace classroom activities with game situations (...) the idea will always be to consider them as another possibility for exercising or stimulating the construction of concepts and notions that are also required for carrying out school tasks." (Petty,1995:p.11)

We mustn't forget that any activity used for educational purposes must be well planned. It's important that the teacher knows the game well, as well as the objectives he or she wants to achieve by the end of its application, so that the lesson doesn't become just entertainment without the acquisition of knowledge.

CHAPTER 4

CLASSIFYING GAMES FOR TEACHING MATHS

There are various classifications, categories and types of educational games that exist in maths teaching, depending on the author. Here we will adopt the classification according to Lara (2003), as it suits the purpose of our work, which are: construction games, training games, deepening games and strategic games.

Construction games are those that bring new and unknown information to the student. Knowledge is acquired through the manipulation of materials or questions and answers so that the student is instigated to want to acquire new knowledge and thus solve the new situation proposed by the game. This type of game requires more from the teacher, both in the design and execution. This is because the student's knowledge is limited and they need to be instructed by the teacher, who acts as a collaborator and guide for the student's acquisition of new knowledge and information. The student will take the initiative to lead this process of knowledge construction. The aim of this type of game is to get the student to reach more advanced levels of conceptual development. This type of game is part of the Constructivist pedagogical trend.

Training games are those used to consolidate certain knowledge or mathematical thinking, not to memorise it but to generalise it, as well as to increase your level of confidence and familiarise yourself with the new knowledge acquired. This type of game leads to a progression in the development of logical reasoning, increasing the possibilities for action and intervention in solving new and even existing problems. In addition, this type of game serves as feedback so that the teacher can see if the student has achieved the desired understanding of certain content, especially those passive students who interact little in class. It is excellent for the teacher to see the real difficulties that students like these have, and through the game to help them clear up their doubts.

This type of game has an undeniable advantage: it replaces uninteresting lessons with repetitive exercises, which often become mechanised and boring, with a pleasurable and motivating activity where the student becomes active and a builder of their knowledge. I'm not saying that lists of exercises are inappropriate, but the way they are prepared or proposed is sometimes demotivating and tiring, and ends up taking away the student's interest in practising and exercising what they have learnt, and as a result, knowledge is lost.

Deepening Games are used to reinforce knowledge acquired through application. Problem-solving is a very convenient activity for students to understand the usefulness of the content, and this type of activity can take the form of games. Games of this type can be developed at different levels, as knowledge deepens.

Finally, Strategic Games are those that allow students to develop strategies so that they can act as players and solve the situations that the game creates. In this type of game, the student is an investigator, who needs to develop systemic thinking, create hypotheses and develop competences and skills to reach the answers they need to solve the proposed problems. The student has to anticipate the move, see the existing possibilities, study the hypotheses and predict events that may or may not interfere with their own or other players' future moves. It's ideal for students to correct possible errors in their strategies and further develop their logical reasoning, which is essential for learning maths.

The PCN's emphasise the importance of strategy games for teaching maths:

> "In strategy games (the search for procedures to win), the starting point is the realisation of practical examples (and not the repetition of procedural models created by others) that lead to the development of specific problem-solving skills and the typical modes of mathematical thinking." (MEC, 1998, p.47).

In strategy games, each player needs to anticipate their moves, visualise what their actions might lead to, or what the other player will do, in order to plan strategically and beat them. In short, strategic games are ideal for developing the skills and competences needed to build mathematical knowledge in the student's cognitive structure.

As we've already mentioned, there are countless classifications of games, each author highlighting a type of criterion to be followed and analysed. But there are some characteristics that are common to all games, according to Flemming and Collaço de Mello (2003):

- Voluntary activity;
- Rules;
- Time;
- Space;
- Material resources.

These characteristics make us realise that the game has to be something free, voluntary and motivating that can be interrupted at any time if necessary. And that the rules can be pre-established or even modified during the game, if all the players agree. Time and space are fundamental characteristics, as they must be appropriate and agreed upon before or during the game, so that the student learns to deal with limits. In a game, there may or may not be concrete material resources. What matters in a game, with the exception of games of chance, is establishing strategies for executing moves and evaluating their effectiveness in terms of the results obtained; in short, the game is not meant to be something passive and mechanised without any meaning for the player.

A very important feature of the game, which contributes significantly to discipline and the social life of the student, is the rules. Games with rules contribute to a relationship of mutual respect between teacher-student and student-student, enabling collective learning where one learns from the other and often takes the other as a reference to try to overcome them. It's learning that losing or winning is relative.

According to Piaget (1978), the rules are the most important part of the structure of the game, they must be mutually respected and modified when necessary. They are used to organise the game collectively and through them the student stops being egocentric and becomes social. The rule establishes what can and cannot be done in the game, limiting the opponent's actions. Moura (1995) describes the social relationships that exist in a game with rules:

"In rule games, the players are not just next to each other, but

> 'together'. The relationships between them are made explicit by the rules of the game. The content and dynamics of the game not only determine the child's relationship with the object, but also their relationships with other participants in the game. (...) Thus, the game of rules enables the development of the child's social relationships." (MOURA, 1995, p.26)

Considering the Game in its pedagogical aspect, we can see that it is an ally for the teacher, as it facilitates the transposition of mathematical language, which is difficult to assimilate, and develops the student's ability to think, act, understand, analyse, reflect, create hypotheses, test them and evaluate them, individually or collectively.

4.1 MATHEMATICAL GAMES IN PRIMARY EDUCATION

When we talk about games, we often think of children and pre-adolescents, but the fact is that games can be used at all stages of our lives. Most of the work on mathematical games is aimed at pre-school and primary school students, but few theorists and researchers have documented research into the importance of this tool for secondary school students, and specifically in the teaching of maths. For this reason, we decided to report on the importance of using games at all stages of basic education and what competences and skills mathematical games can develop in students at different educational levels.

Certainly the school stage that uses games the least in maths lessons is secondary school. Because students at this level are in the transition phase from adolescence to adulthood, many believe that play is part of the world of children, but these thoughts are mistaken. Grando (2000), in his research, found that human beings need the presence of play throughout their lives. An example of this is the existence of games for all age groups and of various modalities, depending on the intellectual level of each person.

> "[...] Man's need to develop playful activities, that is, activities whose

_m is the pleasure that the activity itself can offer, determines the creation of different games and plays. This need is not minimised or modified according to the age of the individual. Engaging in playful activities is a necessity for people at any time in their lives." (GRANDO, 2000, p. 1)

Games are a constant at every stage of our lives, including old age, as a way of passing the time and keeping our minds active. It's common when we have free time or are resting from our work to look for something playful to amuse ourselves, so why not combine the useful with the pleasant: learning by playing in the classroom.

Games are part of human evolution, maturation and learning. It's much easier to learn by playing, because the game itself has components that arouse the learner's interest, making them an active part of the process.

Many people don't use games in high school maths lessons because of the false belief that playfulness can compromise the seriousness that is required of students, who are transitioning to adulthood, and the subject. The education system spreads the idea that maths is a serious subject and that the use of games in maths lessons can compromise this seriousness, as it provides a cheerful and fun environment.

The PCN's suggest that maths should be taught in a diversified way, through attractive and creative games or problem situations that enable students to find strategies that favour mathematical learning. And it emphasises special attention to this subject in secondary school,

"Maths, due to its universality of quantification and expression as a language, therefore occupies a unique position. In secondary school, when a more elaborate construction of the sciences becomes essential, mathematical instruments are especially important." (MEC, 1999, p.211)

In addition, the PCN's advocate the use of games to teach maths, believing that they contribute to the formation of attitudes in students that are essential for developing mathematical thinking and social relationships. Games help to develop strategies, critical thinking and deductive reasoning, which are essential for developing logical-mathematical intelligence.

In short, mathematical games can and should be developed in primary school maths classes, because playfulness and knowledge have no limits or age groups. We are proposing this work in order to verify the contributions that mathematical games can make to primary school students, especially in upper primary and secondary schools, where games are less commonly used.

4.2 GAMES AS TEACHING-LEARNING STRATEGIES: THE THE ROLE OF THE TEACHER AND THE STUDENT.

Working with games requires a lot of dedication, organisation, planning and commitment on the part of the teacher, whose job it is to mediate between knowledge and the student. They must take great care to ensure that lessons don't become a mess, getting away from the main focus and developing meaningful and enjoyable learning. In order for this disorder not to happen, teachers need to know the game well, define the objectives they want to achieve and, above all, guide students towards intellectual development.

It is up to the teacher to create a favourable environment for teaching and learning, to be flexible, subject to the opinions and ideas of the students themselves; to provide an interaction between teacher-student and student-student so that there is also an exchange of experience; to adapt the material to the class and to plan so that at the end of this recreational activity the student has evolved mentally and not just practised an activity to pass the time.

When the game is used as a teaching-learning strategy, the teacher must assume a new posture, different from the traditional one where the teacher teaches and the student learns. In its execution, the game allows for a learning practice that gives the student alternatives and freedom to construct their own knowledge and find answers, often unknown to the teacher, to the problem situations provided by the game.

Borin (1995) emphasises that the role of the teacher, faced with this new resource, is to encourage the student to seek "victory", whether or not they know the winning strategy, because it is up to the player to find it. Firstly, the teacher plans the action, setting goals to choose the ideal game to develop the skills and competences of the content being worked on. Secondly, he or she guides and encourages the student, clarifying the importance of the game as a strategy for learning the mathematical content, highlighting not only the game's playful character, but also its educational one.

Vygotsky developed the concept of the zone of proximal development, which is the distance between the child's actual and potential development. This concept gives the teacher the role of outlining what the student is capable of achieving and identifying their state of cognitive development. In other words, it is up to the teacher to realise what the student needs to learn in order to achieve their real development.

The author Lakomy (2008) clearly defines the role of the teacher in the teaching-learning process:

> "With this we can say that learning is a dynamic process that involves the student interacting with the environment and, for this to happen, it is necessary for the teacher to pay due attention to the factors that motivate the student to learn; to understand that the teaching-learning process is a spiral of knowledge, and that each piece of knowledge is the basis or prerequisite for acquiring the next; have a good critical sense when analysing the child's stage of cognitive development in order to determine what capacities he or she has or doesn't have to work on certain content; stimulate the child's social interaction process (LAKOMY, 2008, p. 48).48)

Howard Gardner, based on the studies of Piaget and Vygotsky, created the *theory of multiple intelligences",* which states that we have different capacities or various intelligences to create something, solve problems, create projects and contribute to understanding our cultural context. Based on this, teachers need to understand that each student has potential and it is up to them to encourage them to develop their intelligences.

Reconsidering assessment methods; providing various opportunities to learn the content; encouraging students to consciously show what they are learning; understanding the existence of differences between students - all of these tasks are the responsibility of the teacher who assumes the position of an educator committed to the teaching and learning of the student. The teacher must encourage autonomy, responsibility and interaction between students so that they can acquire new knowledge.

According to GRANDO (2000), we can't fail to mention the advantages and disadvantages for teachers and students that games can provide when used as a teaching strategy in class, as can be seen in table 4.2.1.

ADVANTAGES	DISADVANTAGES
- Fixing content;	- Misuse of the game (playing
• Introduces and develops concepts; • Develops logical thinking; • Learn to make decisions and evaluate them; • Develop strategies; • Interdisciplinarity; • Meaning of concepts; • They are reusable; • It allows individual (or group) observation and analysis of the student's difficulties; • Correction of any errors; • Facilitates the teaching-learning process; • It develops creativity, competitiveness, critical thinking and the pleasure of learning; • Strengthen or recover student skills; • Diagnosing learning.	without a goal); • Time (requires more time to develop); • Loss of playfulness (when there is too much teacher interference); • False conceptions (thinking that the game will teach everything); • Coercion from the teacher (forcing the student to play); • Inaccessibility to teaching games (lack of subsidised materials for teachers to work with).

Table 4.2.1 Advantages and Disadvantages of using games in the classroom

The student is the protagonist of the educational process. They are the centre of the teaching-learning process, actively constructing their learning, developing competences and skills that trigger their intellectual, affective, linguistic, cognitive, social, ethical and motor development.

In order to play a maths game, the student is responsible for knowing the minimum necessary basis of the related content, so that they can expand and consolidate this knowledge through this enjoyable activity. They need to reflect on their actions, evaluate them and devise new strategies to succeed, in other words, to correct their mistakes. Students need to be open to new information and want to absorb it, otherwise the teacher's effort and dedication will be worthless. It's the student's interest in learning that has to start, so that the teacher can mobilise their efforts and make this educational process a reality.

The game allows students to exchange experiences with each other, to act on behalf of others and to respect their classmates' point of view. Students go from being passive to active, building their knowledge and no longer listening to explanations that often seem pointless.

We can therefore conclude that the role of the student during the game is to participate and get involved, being an active subject contributing to their own learning, only in this way will they benefit from the advantages that this teaching strategy has to offer.

4.3 PLANNING THE USE OF GAMES IN THE CLASSROOM

As we said in Chapter 2, any content can be transmitted and understood by any student, as long as it is organised and structured in a way that suits their intellectual structure. But for this organisation to take place, "planning" is necessary. This is also the case when teachers want to use mathematical games as a methodology in their lessons, they need to plan the actions that will take place.

Teachers need to learn to plan. If they don't plan their day, they don't plan their week, they don't plan their month, they don't plan their year, they don't plan their life. This makes it increasingly impossible for them to achieve the ultimate goal of education, which is the student's learning.

Planning has everything to do with the goal you want to achieve. Those who plan know what they want and how they want to achieve it. When teachers want to use games as a teaching strategy in maths lessons, they need to know what they want to achieve by developing the game and how they want to achieve it. They need to know what content can be worked on with that game and what skills and competences this activity can provide to the student.

You don't just have to choose the game, you have to know if that game is ideal for achieving the desired teaching objective, which is why it's important for teachers to plan before introducing any activity. Planning requires discipline and vision. Teachers need to be aware that the game is an activity that requires time and that it may not produce immediate results.

When planning, the teacher needs to put themselves in the student's shoes, see what difficulties they may have in relation to the content they are working on and see how the game can help to resolve these difficulties. It is essential that the teacher knows the game in detail (they need to play the game before applying it), so that they can clarify any doubts the students may have as the game progresses and can also ask some questions that induce the student to develop their mind more and more in order to add new information that will lead them to meaningful learning.

Good lesson planning requires the teacher to be clear about the following questions, and to check whether the activity (or game) they are going to introduce will meet their expectations in relation to their proposed objectives for that lesson. These are

- Who is the student?
- What goal do I want to achieve with the topic I'm going to develop?
- How can you meet the needs and requirements of each student?
- How to start, develop and finish the lesson?
- Do I know a suitable activity (game)?
- Do you need to make adaptations (to the activity)?
- How to apply it? And when to apply it?
- What materials are needed?

When working with games as a pedagogical tool in the classroom, it is essential for the teacher to plan the actions that will be carried out and the objectives that will be achieved. If this organisation of ideas doesn't take place, the application of this activity will only be of a playful, recreational nature, and the lesson could turn into a mess (disorder) where it will frustrate the teacher's expectations and the student won't add any significant learning, as they won't perceive any relationship between the game and the mathematical content.

In short, when working with games, teachers need to plan their lessons well so that this activity fulfils both its playful and educational character. Planning is part of the teaching-learning process, as it defines what you want and how you want to achieve it. Taking care when choosing the game and how to present it is also part of this stage, as the way it is presented will affect whether or not students are motivated to take part in this activity and consequently want to learn the content presented.

CHAPTER 5

AN EXPERIENCE USING GAMES IN PRIMARY EDUCATION

[a]We are now going to report on an experience we had using the game "roulette of integers", with students from the 6th grade/7th year of the Mirtes Demes School Unit in the city of Floriano-PI, in 2011. This experience is reported in more detail in the article ***"A proposal for teaching and learning with mathematical games: the roulette of integers".***

With the aim of reducing students' difficulties in relation to the content of operations with integers, we were curious to use the game "roulette of integers" as a form of intervention to help and try to create a more favourable environment for producing more meaningful and enjoyable learning, and also to check whether the theories on the use of games for teaching were true.

This experiment was carried out in three stages, starting with a pre-test for a diagnostic assessment of reality, which was done by applying an activity involving situations and problems with the content in question. This survey was used to check the students' knowledge of operations with whole numbers.

Next came the application of the chosen game, where the students were observed at every stage of the game's development. At this stage, the students were analysed quantitatively and qualitatively. Quantitatively, we looked at the number of points each student had scored and qualitatively, at their behaviour, participation, concentration, motivation and interest.

In the last stage, the post-test was carried out, consisting of problem situations similar to those in the first activity so that the students could be assessed again with the aim of realising the game's contribution to their learning.

5.1 GAME DESCRIPTION

INTEGER ROULETTE

Roulette model

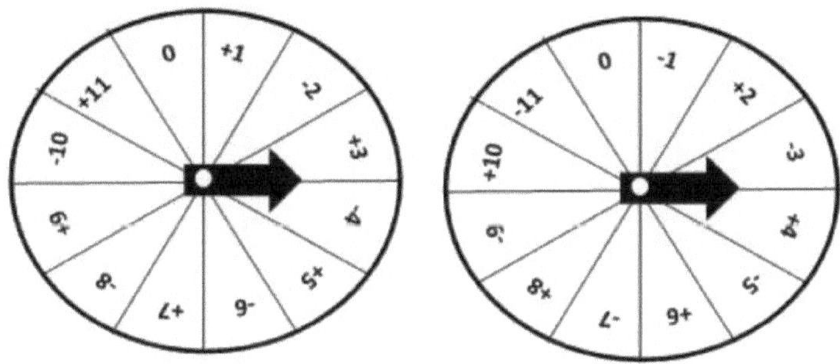

Data model

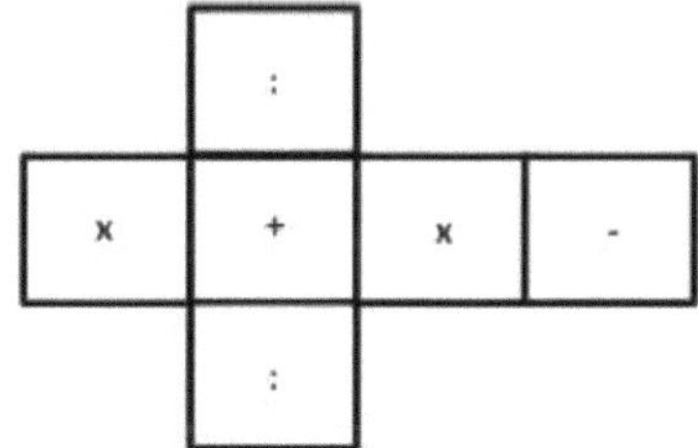

Source: *LARA, Isabel Cristina Machado de.* ***Jogando com a Matemática*** *de* **5ª a 8a série**. *1ª ed. São Paulo: Rêspel, 2003.*

Objectives: For the student to be able to:

R Solve operations with integers;

-Differentiate the rules of constructed signs;

-Develop your ability to make mental calculations;

-Fix mathematical content;

-Create resolution strategies.

Prerequisites:

- Addition, subtraction, multiplication and division with whole numbers and decimals.

No. of players: at least 2 players

Materials:

- 1 dice with signs - 2 roulette wheels - writing materials

How to Play:

The first player, chosen by the group, spins the two roulette wheels and throws the dice. They perform the indicated operation between the two numbers drawn and only record the result in a table. If they get it right, they score a point. The other players proceed in the same way up to a number of rounds set in advance. At the end of the game, the players add up all the results and whoever gets the highest score scores 5 points. The player with the most points at the end of the game wins.

Material used to make the game:

The dice can be made out of poster paper, modelled on a cube. Place 1 addition sign, 1 subtraction sign, 2 division signs and 2 multiplication signs on the faces. The roulettes can be made from poster paper and protected with contact paper.

5.2 ACTIVITY ANALYSES

Initially, the pre-test showed that 100 per cent of the students had difficulties performing operations with whole numbers and associating the content with everyday problems.

In many questions it was noted that the students understood what was being asked, but didn't associate it with mathematical calculation and often confused the operation they were supposed to use in the given question.

Although the questions were quite clear, the students found it very difficult to solve them, as they have difficulty interpreting even the simplest and most obvious questions. This is because they are used to solving lists of exercises, where they just substitute formulae without knowing their applicability,

that do not stimulate their capacity for reasoning, interpretation and creativity, contrary to what was perceived in the application of the game roulette of integers.

In short, the pre-test showed that the students had not learnt the operations with whole numbers in a meaningful way, since many times during the pre-test it was noticeable that many of them were struggling to remember the rules they had "memorised" and ended up confusing them, operating in the wrong way and performing poorly in this activity; at this stage the class average was 3.2.

Starting with the game in question, roulette of the integers, the students showed a lot of enthusiasm and expectations for the activity. All the students were interested, curious and motivated to take part in the lessons because the game was a different way of teaching, without forgetting that learning was taking place in an attractive, fun and challenging way, essential characteristics for better performance in the teaching-learning process.

The "Roulette of the Integers" game was played over five lessons, i.e. one week. During this period, we noticed some positive aspects in the students, such as: attentiveness, participation, curiosity and concentration. Aspects like these are quite difficult to notice in traditional classes, especially those aimed at teaching maths. Attention and concentration are essential for both teaching and learning to take place, because not even a teacher can teach without students paying attention, nor can students learn and develop their reasoning without attention and concentration. We also know that curiosity and participation are fundamental to meaningful learning, since students who are curious and participate in class perform better in their subjects.

Roulette of integers was a sensation in class, the students didn't see the time passing, they always wanted to take part, even when it wasn't their turn, and they didn't even realise that they were acquiring mathematical knowledge at the same time as having fun. Everyone wanted to get the operations proposed by the game right, and when a student made a mistake, other classmates would point out the error and tell them the correct answer, in other words, at all times we could see that knowledge was being produced even through the mistakes they made.

At the end of the period in which the game was used, we could see a significant improvement in the students' performance, as they gained concentration, developed the ability to make mental calculations, which is fundamental for mathematical reasoning, and felt comfortable learning, so that lessons that used to be torture ended up being enjoyable and fun. As far as the quantitative aspects are concerned, we can see that there has been a significant improvement in the students' performance; at this stage the class average rose significantly to 7.5.

At the end of the "Roulette of the Integers" game, a post-test was administered to find out whether or not the game had contributed to the students' learning in quantitative terms, because qualitatively it was noted that there had been a significant improvement. The test applied was similar to the pre-test, with questions of the same level, so that a comparative analysis could be made of the similar questions solved before and after the game, and only then conclude to what extent the game "Roulette of the integers" helped in the learning of operations with integers.

By evaluating the students' tests, it was possible to see that the errors they had previously made had been reduced by around 50 per cent. These were related to sign operations and applications in problems involving operations with integers. Finally, considering the period during which the game was applied and the results obtained from the pre-test, we can see that there was a significant improvement and at this stage the students' average was 6.5.

Finally, it was found that the game "Integer Roulette" contributes to significant learning in terms of operations with integers. Because the average performance of the students before the game was 32%, during the game there was a significant increase of more than 100%, bringing the average to 75% and after the game this significant increase was still above 100%, bringing the average to 65%, with only a decrease of 7.5%, a decrease that is considered insignificant given that learning occurs continuously and at different times for each student. The results of the test confirmed that the students' performance improved significantly and that the game "Roulette of the Integers" really did help to improve the students' learning of operations with integers.

5.3 FINAL CONSIDERATIONS

Based on the analyses and discussions, we can see that the game "Roulette of the Integers" makes a qualitative and quantitative contribution to the learning of operations with integers by 7th grade students at the Mirtes Demes School Unit in the city of Floriano-PI.

Qualitatively, it was noted that students in classes using the "roulette of integers" game had more attention, concentration and participation, developed strategies for solving problems, learnt from their own mistakes and those of their classmates, and there was a reduction in their blockages with maths, because learning was taking place in a fun and enjoyable way. With this, we can say that the game promotes favourable conditions for the teaching-learning process. So the quantitative performance could only be a consequence of this whole set of favourable situations for teaching mathematics.

Games in maths education have a material teaching character from the moment they are used to promote learning. It can therefore be very useful in the process of teaching maths in order to break down the barriers between students and the subject, the vast majority of whom see maths as an invincible seven-headed beast and this ends up blocking their performance. This is what we saw with the "Roulette of the Integers" game, which significantly increased the performance of students who previously had a below-average performance and managed to achieve a significant increase after applying the game.

It was observed that the game "Roulette of the Integers" not only contributed to the evolution of the students' learning, but also aroused their interest in mathematics, since the use of the game proposed activities that stimulated active participation, aroused attention and interest, challenged reasoning and helped the students to reflect on their actions and analyse them in order to create their own solution strategies. This allows students to be the builders of their own knowledge.

The results of the tests confirmed that the game "Roulette of the Integers" helped to improve the students' performance. This intervention not only contributed in numerical terms, but also in behavioural and attitudinal terms, improving their solving strategies, which is fundamental for teaching mathematics.

CHAPTER 6

AN EXPERIENCE USING GAMES IN SECONDARY EDUCATION

°We will now report on an experience we had using a "mathematical trail" game with students from the 4th year of the Integrated Technical Course in Oral Health at the State Centre for Professional Education (CEEP) in the city of Floriano-PI, in October 2014. This experience is reported in more detail in the master's thesis entitled ***"A proposal for teaching and learning with mathematical games in secondary education".***

Unlike the experiences we had with maths games in primary school, where we worked with the class in which we were developing the games, the experience with games in secondary school was carried out with a class in which we weren't the head teacher, due to the fact that I wasn't teaching in secondary school at the time.

The proposal for this work was developed in stages, as follows: 1) Search for bibliographic materials related to the theme "games in the teaching of maths", as well as a school for the development of the research. 2) Study of the mathematical content being taught in textbooks and choice and manufacture of the materials needed for the chosen game.3) Implementation of the workshop. 4) Analysing and discussing the results.

In the first stage, we found it very difficult to locate bibliographic materials related to games in the teaching of maths in secondary schools, which was our focus, because in libraries, bookshops and even on book and educational websites, these materials are few and far between. In our reality, unfortunately, there is still little published work on the use of games as a methodology in the maths teaching-learning process.

In the search for a school that would agree to carry out our project, because we weren't teaching the target audience, a major difficulty was encountered: the availability of time to carry out the proposed activities. Most of the teachers argued that their schedules were full and that they didn't have time in their workload. Even when they explained the scale of our project and that it could be contributing to their professional curriculum by adding value to their teaching practice, we could see that they were concerned with explaining as much content as possible that was included in the curriculum, rather than the quality and development of student learning.

When we found a teacher who understood the scale and importance of our research, we set about studying and analysing the mathematical content that was being worked on in the classroom, Combinatorics and Probability. We researched various textbooks that dealt with this content in order to find a language in which students could understand its real importance and significance for their everyday lives. We then went in search of a game that would give greater meaning to the content covered and chose to work with the "Maths Trail" game, as it is a game that can be used to train the student, fix concepts and deepen knowledge through problem solving. At this stage, we selected the questions that would make up the game and made the materials needed for it.

The workshop was the culmination of our work, as it was where we had direct contact with the subjects of the investigation and were able to verify and analyse whether theories and studies on games in the teaching of mathematics can in fact contribute to the teaching-learning process. The workshop began as soon as the head teacher had covered the expected content, Combinatorics and Probability, including the application and correction of exercises on the subject. In the first phase of the workshop, we carried out a review to diagnose the knowledge acquired by the students and clarify any doubts related to the content, and even to familiarise ourselves with the students before applying the "Mathematical Trail" game. The game was then applied so that we could closely analyse the real contributions that this teaching strategy makes to the development of student learning.

During the review, we used a multimedia device to help present a review of the content covered and some problem situations that clarified the applicability and usefulness of that content in everyday life. This phase of the workshop saw considerable participation from some of the students, who took advantage of the revision to ask questions about the subject. This participation was fundamental for us to get to know the students, because at this stage we were able to diagnose their behaviour, difficulties and the knowledge they had already acquired. We tried to carry out a participatory revision in which the students were encouraged to solve the problems proposed during the lesson. We can say that seventy per cent of the class took part, which we consider to be an excellent rapport, given that it was our first contact with the class.

For reasons of time, we opted to use just one game with characteristics that would develop the skills and competences inherent in teaching Combinatorics and Probability, involving rules, luck, reasoning, problems and information that would broaden each player's knowledge. However, we suggest using more than one game to develop a subject, because the more activities and applications, the better for the student's learning and intellectual development.

6.1 DESCRIPTION OF THE GAME

MATHS TRAIL

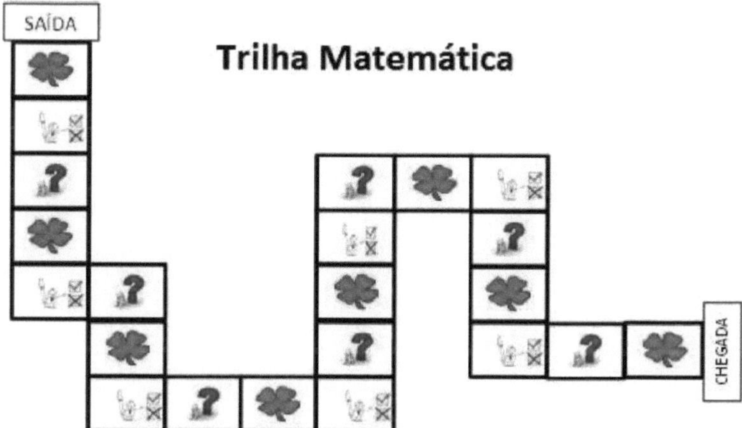

Content: Can be used for any mathematical content.

No. of players: the whole class divided into groups of three or three large groups (if teamwork is preferred).

Materials:

- Track;

- Data;
- Markers;
- Paper, pen or pencil;
- Question bank (referring to the content worked on).

Note: All the material can be bought or made by the teacher. The material used to make the game was wood paper, white glue, paintbrushes, dice (which can be bought or made), pawns (to represent each player) and printed materials (which can also be drawn or written on, depending on the creativity of each teacher).

How to play: Each player places their marker in the starting position and decides who starts the game. The first player rolls the dice and moves forward the number of squares indicated by the dice and carries out the proposed activity, for example, if it's a problem or a true or false question to be solved, they choose a number from 1 to 20 (which is the number of problems and V or F questions in the question bank), and if they answer correctly they win the right to play the next round, otherwise they don't play for one round. The students can solve the questions using pen and paper, and they can also ask the teacher any questions they may have, but under no circumstances should the teacher solve the questions for the student, their role is simply to instruct. And so everyone throws the dice on their turn and solves what is asked of them. If the player stops on a lucky or unlucky square, they will move forward, back, stay a round without playing or play again, depending on another throw of the dice, each face of which will indicate what will happen according to the game's legend, which is at the teacher's discretion. The game ends as soon as one player reaches the finish line.

6.1 ACTIVITY ANALYSES

The "Maths Track" game is a rule-based game that can be classified as training and in-depth. It has true or false questions and problems related to the subject in its squares, as well as lucky or unlucky squares.

To play the game, the students were organised into trios and each trio was accompanied by a monitor to answer questions and observe their behaviour and performance during the game. Each monitor had a bank of listed questions, some true or false and others problem situations.

In the question bank there were problems at different levels, from the easiest to the highest, which is why students chose a number without knowing the level of the question. This aroused their curiosity and interest, and they became part of the group, because if a colleague chose an easy question, everyone else would solve it and make comments like "that's easy", "everyone knows how to do that one", or "even I know how to do that one". Often they even wanted to solve it for their colleague, but as it was a competitive game, they would wait for their turn and hope that they would choose questions that they could solve and progress further and further in the game. When the question was at a higher level, we could see the interest of other colleagues in trying to solve it to test their knowledge. There was a noticeable rapport between the players and motivation in the faces and attitudes of each student.

On the squares that were lucky or unlucky, the student was not left at a loss for what to do, because even on these squares he was directed to move on or back to other squares to solve problems, luck was just a matter of moving forwards or backwards in the game. Since the main objective was to train and deepen knowledge, we couldn't leave the student without exercising their knowledge in each move.

We realised that the classroom had become a lighter, more interesting environment, where the subjects were active participants. In the groups formed, you could see joy on the students' faces, enthusiasm and a desire to learn, to get it right. The students tried to help each other by giving tips, because when they realised that they knew how to solve a colleague's problem, and the colleague had difficulty solving it, they wanted to help and teach them how to solve it. When the students had difficulties or doubts about solving the problem, they would ask the teacher who was monitoring them, so that they could put together their strategies and solve the problem correctly.

In the game, it was possible for the questions that a player got wrong to be chosen again by another player on their turn. This allowed the player who didn't get it right to learn from their colleague and understand what they had got wrong. Thus, learning was both individual and collective, as there was a constant exchange of information between student-student and teacher-student.

The rules of the game stated that the game would end as soon as one player reached the finish line, but when this happened the other students wanted it to continue, because they wanted to continue exercising and proposed that everyone should reach the finish line, regardless of whether they arrived first or last, the important thing would be for everyone to complete the journey and acquire more knowledge by solving more questions. Even those who reached the finish line first kept watching the others who continued, giving them tips, cheering them on and testing their knowledge of the content. At the end of the day, they asked to solve the questions that none of them had managed to solve, and some even asked the head teacher to solve some of the more complex questions at another time, due to time constraints.

Finally, we saw interaction, enthusiasm, motivation, commitment, joy, relaxation, pleasure, exchange of information, development and, most importantly, learning. At the end of the workshop, we heard from some students about what they thought of this type of practice in the classroom, games as a strategy for teaching maths, and we had positive responses.

The students found it very interesting, because they got out of their routine, which for them "is boring". They had fun and didn't even realise they were studying maths. One student went so far as to say that it was a way of attracting their attention so that they could learn this subject, because most students aren't big fans of maths and aren't very interested in doing lists of exercises or studying them at home, because they don't see how they apply to their reality. But with the game they were able to see the contextualisation of the content in the problems and how it has to do with reality and can be interesting, depending on how it is presented. One student went so far as to say "The use of games should be compulsory in maths lessons!".

When we asked the students what should be done to improve the quality of teaching and attract their attention in maths lessons, they said that every lesson should have a different, dynamic activity that motivates and arouses their interest. Indirectly, they need to feel active, useful and not just puppets in the teaching-learning process.

The game is an activity that turns students from passive to active subjects, and this strategy can reduce the failure rate in this subject, because everyone learns in a much more interesting way. And the introduction of this didactic-pedagogical tool into maths lessons can make them more attractive and no pupil _would_ be frowning and begging God for the lesson to end soon.

We asked the head teacher what her opinion was on using games to help in maths lessons and whether she had ever used this methodology. She replied that it was very good to work with games in lessons to teach maths. And that she had already used concrete and manipulative materials, simulating bingos and even draws when she started the subject of probability, but she had never worked with it in the way it was done in the workshop. Unfortunately, the CEEP (State Centre for Professional Education) has a brand new, state-of-the-art maths laboratory, but for bureaucratic reasons it has never been used, all in plastic, due to the fact that the technicians trained to set it up and train the teachers (or monitors) to use it properly are from São Paulo and so far nothing had been done. This is a waste, as we believe that this laboratory has many marvellous educational resources (games, manipulative materials, computer resources), which could be used to help teachers and contribute to meaningful student learning.

The head teacher also reported that there is still a lack of training for teachers to develop innovative methodologies in the classroom. Time was another factor mentioned: "They throw a lot of content at us during the school year, and we only have two maths lessons a week! We have to make do in thirty. Teachers really do what they can, and sometimes what they can't, when selecting content to teach, because the time that is calculated doesn't take into account the fact that a classroom is heterogeneous, and that each student has their own learning time. We mustn't forget another factor that often prevents teachers from carrying out different activities: their time to plan them. This is due to the low professional value placed on teachers, who often earn very little and are therefore forced to work in several schools in order to live a dignified life with a minimum of comfort.

Talking about the students' behaviour and interest, we asked the head teacher if the game had given them something new, different from their everyday lives, "Absolutely! They were more committed and took an active part. I could see the interest in their eyes, in their attitudes, in the questions and doubts they raised during the workshop. And the feeling of pleasure and fun while the game was being played was evident on each student's face."

But the teacher's task is arduous, and it takes commitment to make a difference. To achieve excellence, you have to make a difference. You can't let difficulties prevent you from achieving the main goal of education: learning. It takes effort and dedication to teach, because in order to achieve objectives, we need to set goals and chase after them. Teachers committed to teaching and learning need to make a difference, proposing methodologies that adapt to the real needs of their students.

To further enrich our research, we also analysed the students quantitatively, with the notes provided by the teacher on the activities she applied, one before the game was applied and one after. The activities applied were similar to the problems in the "Maths Trail" game, which can be found in the appendix of this paper. The class average in the first assessment was 6.2 and in the second assessment it was 7.9, an increase of 27.4 per cent. This percentage increase in marks is quite significant, given that learning takes place continuously and at different times for each student, without forgetting to mention that the workshop was held in a short space of time.

This experience with the "Maths Trail" game was very fruitful, as we realised that there was a greater rapport and desire on the part of the students to learn maths. They were motivated to become active subjects and builders of their own knowledge. The game as a didactic-pedagogical tool is a valuable instrument for reducing some of the blockages between students and maths. The students' level of participation, performance, motivation, relaxation, learning, among other benefits, was evident with the implementation of the "Maths Trail" game as a different proposal to assist in the teaching-learning process of maths teaching.

6.2 FINAL CONSIDERATIONS

From the observations and analyses made with the students of the 4th year of the Integrated High School to the Oral Health Technician at CEEP, we can see that the game "Trilha Matemática", as a teaching strategy, does contribute, qualitatively and quantitatively, in a significant way to a more enjoyable and engaging learning of mathematical concepts related to Combinatorics and Probability, developing in the research subjects competences and skills inherent in the teaching of Mathematics.

We observed that during the development of the "Maths Trail" game, the students increased their level of concentration, interest, participation and motivation. They developed reproduction, reflection and connection skills when faced with the problem situations that the game provided. Learning was taking place in a fun, enjoyable and spontaneous way. There was a reduction in blockages on the part of some students who said they hated maths and a healthy interaction between the students. The game provided a favourable environment for teaching mathematics.

When games are used to teach maths, they can help teachers in the teaching process, becoming an ally to ease the obstacles between students and mathematical content, improving student performance in this subject. This is what we observed with the application of the "Maths Trail" game, which aroused the students' interest in wanting to learn this dazzling science that is so useful in human life.

Using games in the classroom without planning and support to explore their possibilities and effects on the maths teaching-learning process can turn them into just another recreational activity that adds no value or meaning to the students' lives in order to develop their mathematical thinking. Most teachers still use games in the classroom just to fulfil a motivational role without exploring any kind of reflection on the mathematical structures linked to the action of the game.

Our intention was not to analyse the game itself, "Trilha Matemática", but rather the attitudes, behaviours and skills that this activity can promote in order to help the teaching process, facilitating the development of students' cognitive structures. The aim is to see the game as a playful instrument that enables the educational process. The game takes on a pedagogical character when used as an intervention tool in teaching, playing a didactic-pedagogical role to remedy some of the difficulties encountered by the student in the teaching-learning process.

During the intervention period, it was observed that the game not only contributed to the development of the students' learning, but also aroused their interest in maths. The "Maths Trail" game stimulated active participation, aroused attention, challenged reasoning and contributed to the students' analysis and reflection on their actions, making them create their own resolution strategies. Thus, it was evident that the game rescued and applied mathematical concepts previously constructed by the student.

The social aspects of the game were another important and interesting factor to mention. The students learnt to respect the limits and rules that were outlined in the game, promoting the confrontation of ideas, discussions and formulation of strategies. The interaction provided an exchange of ideas, knowledge, experiences and collective learning.

We can see that the use of games as a teaching strategy in maths lessons can be very convenient, but when applied to large classes, it is necessary to have monitors with the teacher to help the activity run smoothly and avoid possible disorder in the class so that the real objective doesn't get out of focus.

The teacher's planning is essential when developing this type of activity, as it is important that the objectives are clear, so that the intervention is made at the right time and in a way that provides meaningful learning for the student. Activities carried out at random don't bring any meaning to the student and even demotivate them, because every human being needs to understand the reason for all the actions they carry out in order to become interested and involved.

Finally, considering all our experience and the facts mentioned here, we can conclude that the game, when used as a strategy for teaching mathematics, not only contributed in numerical terms, but also in behavioural and attitudinal terms for the research subjects. And it is only considered teaching material when its use promotes learning.

In view of the above, we can state that our research was intended to contribute to a reflection on the teacher's didactic-pedagogical action and to improve the teaching of maths by suggesting a teaching strategy that reduces the existing blockages between the student and the object of knowledge, maths. We understand that this work does not end here, as we intend to continue this investigation on other occasions, carrying out new studies and research that prove or rectify the information in this research.

CHAPTER 7

REFERENCES

ALVES, Eva Maria Siqueira. Playfulness and the teaching of maths: a possible practice. 5.ed. Campinas: Papirus, 2009.

ANTUNES, Celso. Multiple intelligences and their games: logical-mathematical intelligence, vol.6. Petrópolis, RJ: vozes, 2006.

BORIN, Julia. Games and problem solving: a strategy for maths classes. São Paulo: IME, 2002.p. 09.

D'AMBRÓSIO, U. Matemática, ensino e educação: uma proposta global. Temas & debates, São Paulo, v.4, n.3, 1991. p 1-15.

FIORENTINI, Dario; MIORIN, Maria Ângela. A reflection on the use of concrete materials and games in maths teaching. Boletim SBEM - SP, São Paulo, v. 4, n. 7, p. 5-10, jul./aug. 1990.

FLEMMING, Diva Marília; COLLAÇO DE MELLO, Ana Cláudia. Creativity Didactic Games. São José: Saint - Germain, 2003.

GRANDO, Regina Célia. The game and its methodological possibilities in the maths teaching-learning process. 1995. Dissertation (Master's in Education) - State University of Campinas. Campinas, 1995.

_______ . Mathematical knowledge and the use of games in the classroom. 2000. Doctoral thesis. Faculty of Education - State University of Campinas. Campinas, 2000.

KAMMI, C.; DECLARK, G. Reinventing arithmetic: implications of Piaget's theory. São Paulo, Campinas: Papirus, 1992.

KISHIMOTO, T. M. (org.). Game, toy, play and education. São Paulo: Cortez, 2001.

LAKOMY, Ana Maria. Cognitive Theories of Learning. 2. ed.Curitiba: Ibpex, 2008.

LARA, Isabel Cristina Machado de. Jogando com a Matemática de 5ª a 8ª série. 1ª ed. - São Paulo: Rêspel, 2003.

MEC - Ministry of Education - Department of Basic Education - PCN's National Curriculum Parameters. Brasília: MEC/SEF, 1998 & 1999.

MOURA, A. R. L. A Medida e a Criança Pré-Escolar. Campinas-SP, 1995. Doctoral Thesis. Faculty of Education, UNICAMP.

MURCIA, Juan Antonio Moreno (ed.) Learning through Play.
Trad. Valério Campos. Porto Alegre: Artmed, 2005.

PETTY, A. L. S. Essay on the Pedagogical Value of Rules Games: a constructivist perspective. São Paulo - SP, 1995. 133p. Master's dissertation. Institute of Psychology, USP.

PIAGET,J. Where is Education going? 3. ed. Translated by Ivette Braga. Rio de Janeiro: José Olympio. 1975. 80p.

_______ . The Formation of the Symbol in the Child: Imitation, Play and Dream, Image and Representation. 3. ed. Translated by Álvaro Cabral and Christiano Monteiro Oiticica. Rio de Janeiro: Zahar, 1978. 370p.

Printed by Books on Demand GmbH, Norderstedt / Germany